APPENDIX

TO THE

REPORT OF THE GEOLOGICAL SURVEY

OF

NORTH CAROLINA,

1873;

BEING A BRIEF ABSTRACT OF THAT REPORT

AND

A GENERAL DESCRIPTION OF THE STATE,

GEOGRAPHICAL, GEOLOGICAL, CLIMATIC AND AGRICULTURAL.

W. C. KERR, State Geologist.

RALEIGH, N. C.:
STONE & UZZELL, STATE PRINTERS AND BINDERS.
1873.

NORTH CAROLINA.

GEOGRAPHICAL DESCRIPTION.

SITUATION.

North Carolina is situated on the Atlantic slope of the great Appalachian chain of mountains, which traverses the North American continent in a Northeast and Southwest direction from Canada to Georgia.

Its territory lies between the parrellels 34° and 36½° North latitude, midway between New York and the Gulf of Mexico, *the latitude being that of Southern Spain*, and between the meridians 75½° and 84½° West longitude.

EXTENT.

The State has a coast line of more than 200 miles, and a length, from east to west, of 485 miles.

Its *area* is 50,704 square miles, which is a little greater than that of New York, *and almost exactly that of England.*

PHYSICAL GEOGRAPHY.

The State is naturally divided into three distinct and well characterized regions, east, middle and west. The latter, or

The Western Division is quite mountainous, and is about one-fifth of the area of the State, (10,000 square miles,) and consists for the most part of a narrow plateau, whose elevation is 2,000 to 3,000 feet above the sea, lying in a Northeast and Southwest direction, between two parallel ranges of the

Appalachians, the Blue Ridge and the Smoky Mountains; its length being more than two hundred miles, and breadth from 30 to 50 miles.

Mountains. This plateau is not only the most elevated region of the United States east of the Mississippi River, but is the culminating region of the Appalachian mountains, and contains its highest peaks, and most massive spurs, the Black Mountain in this State being some 400 feet higher than Mt. Washington, in New Hampshire.

This plateau is traversed also by half a dozen cross chains, which are higher and more massive than the principal ranges above mentioned. Many of these mountains are more than 6,000 feet, and quite a number reach nearly 7,000; the Black is 6,710, and Clingman's 6,660 feet; Mt. Washington, N. H., 6,288. The plateau is subdivided therefore into a number of smaller plateaus or basins, bounded on all sides by mountains, and each having its own independent drainage system. The Blue Ridge, which bounds this plateau eastward, separates it from the

Middle Region of the State, which may also be described as a low plateau, whose western side has an elevation at the foot of the Blue Ridge of 1,000 to 1,200 feet, and is roughened by many spurs of that chain two and three thousand feet high, and many of them 20 and 30 miles long. This region descends very gradually towards the east, preserving an elevation of 600 to 800 feet for 150 miles, constituting the hill country of the State, and having an eastward extent of more than 200 miles, and an area of more than 20,000 square miles.

The Eastern Section, which lies on the seaboard and extends inland 120 to 150 miles, is for the most part comparatively level, or but little rolling and hilly towards the west, and is about equal in area to the last, containing about 20,000 square miles of territory. This region is diversified by many *Sounds, Bays and Lakes,* communicating with its many large navigable rivers and constituting, with the connecting canals, an extensive system of water communication with the eastern and middle sections of the State.

Rivers. There are seven large rivers, flowing east and southeast from the Blue Ridge, through the middle and eastern divisions of the State, besides numerous smaller streams, which furnish indefinite water power through the middle section; and in the eastern, together with the bays and sounds, they give an aggregate of more than 1,000 miles of inland navigation.

West of the Blue Ridge there are seven other large rivers, which flow westward into the Ohio and Mississippi, the largest of these being the great Tennessee, which is larger than the Danube, and is navigable from the western boundary of this State for a thousand miles to the Mississippi.

Railroads. There are more than 1,100 miles of railroad already built, and several hundred more projected, which will be completed in a few years.

Seaports. Wilmington, Beaufort and Newbern are the principal shipping points within the State; and Norfolk, near the northern border, derives a large part of its business from this State.

GEOLOGY.

The geological structure of the State is very simple, the formations being arranged in zones parallel to the dominant mountain system, and to the Atlantic coast, and belonging almost entirely to two systems or ages, the *Primary* and *Tertiary;* the *Secondary* being represented only by two small troughs of Triassic in the middle region, and a few linear outcrops of Cretaceous near the coast. A thin covering of *Quaternary* overlies portions of the Tertiary.

The Primary Rocks, which occupy the western and middle regions, consist of granites, gneisses and schists, of the Laurentian division, with occasional narrow belts of Huronian slates, sandstones, limestones and quartzites; the most extensive of these belts being 20 to 30 miles wide, and lying near the eastern margin of the middle region.

The Tertiary (and Quaternary,) occupies the eastern champaign section, and consists mainly of beds of uncompacted clays, sands and marls, belonging to the lower and middle divisions, (Eocene and Miocene,) which are every where filled with exuviæ and bones of marine animals, constituting an inexhaustible resource of manurial matter.

MINERALS

Are found in great variety and abundance over a large part of the State. Among the more useful and important, are the following: *Marl, Iron, Coal, Peat, Limestone, Gold, Copper, Silver, Lead, Zinc, Mica, Graphite* and *Corundum;* besides *Manganese, Kaolin, Fireclay, Talc, Agalmatolite, Whetstone, Grindstone* and *Millstone grits,* a great variety of building stones, *Serpentine, Marble, Chromic Iron, Barytes, Oil Shales, Buhrstone, Roofing Slates* and several precious stones, as *Diamond, Garnet, Sapphire, Ruby, Beryl* and *Amethyst.*

Marl is found only in the eastern region, but is very abundant in some 25 counties, occurring in extensive superficial beds, which contain all the elements of a complete and permanent fertilizer, an occasional dressing, (once in 15 or 20 years,) being sufficient to render a poor soil permanently productive. This is the most valuable mineral in the State, as it is easily accessible to more than half of the farming lands, and is applicable to all crops.

Iron. The State contains a vast quantity of iron ore of every variety, distributed over a very wide area from the head of navigation on the Roanoke, for example, for nearly four hundred miles westward, to the extreme limit of the State, being found in workable quantities in not less than 30 counties. But a more important fact than the variety or the abundance, or the wide distribution of these ores is the remarkable purity of many of the deposits. Iron has been smelted for a hundred years in the middle and western counties, both in forges and furnaces; but only in quantities sufficient for neighborhood consumption. Much of this iron, though so rudely prepared, is equal to the best Swede, being of course, like that, reduced with charcoal. The most abundant kind of ore is Magnetite, and most of the iron hitherto manufactured in the State has been made of it; but Red Hematite is scarcely less abundant, and Limonite is very common. Two kinds of carbonaceous ore occur in association with the coal, viz: the Scotch Black Band and Ball ore, (calcareous siderite). Many of the beds of the two former ores, Magnetite and Red Hematite, are entirely free from both Sulphur and Phosphorus; some of them contain Maganese, others Titanic Acid, and still others both of these minerals, together with a small percentage of Chromium. Such deposits in such quantities and of such purity, remain undeveloped only because occurring in a region heretofore little accessible, and wholly devoted to agriculture. But their high value for the manufacture of the best kinds of cutlery steel, and for the Bessemer rail is bringing them prominently into notice and demand.

Coal. The coal of this State is of Triassic age, mostly bituminous; is a good gas coal and also well adapted to iron smelting when coked. There are two coal beds, both in the middle region, one on Deep River, mostly in Chatham county, the other on Dan River, (upper waters of the Roanoke,) in Rockingham and Stokes counties. The thickness of the workable seams ranges from 3 to 7½ feet. The outcrops are respectively estimated at about 30 and 40 miles, and the probable breadth at about 3 miles in one case and 1 to 2 in the other. Both of these coal beds are in immediate proximity to some of the most extensive and valuable iron ore deposits in the State and on navigable streams.

Peat exists in very large quantities, (several hundred square miles in area and many feet thick,) in the counties near the seaboard. It is used extensively as a fertilizer by the best farmers, and will some day be of great value for fuel.

Limestone, though not abundant in the State, is found in more than 20 counties; some of them in the eastern, some in the middle, and some in the western region. That in the east is of Eocene age and is a shell conglomerate, valuable both for building purposes and for the manufacture of lime. The limestones of the middle and western regions are of Huronian (Pre-Silurian) age, and are frequently crystalline, and in several counties constitute a very good marble, in Cherokee and Macon especially, where are found several fine varieties and colors of this stone, white, grey, red, flesh-colored and mottled.

Gold is very widely distributed through the older rocks of the middle and western sections, being found in workable quantities in 29 counties. The first gold mines in the United States were found here about 1820, and they were wrought on a very large scale until 1847, yielding many millions of dollars. There has been comparatively little done in these mines since the discovery of the California deposits, although a number of mines are still wrought from Halifax to Cherokee. The mineral is found in various gangues, besides the free gold of the drift or gravel beds; chiefly in quartz, quartzitic slates and conglomerates, talcose slates, felspathic slates, limestone and gneiss. Before the discovery of the California deposits, the largest nugget in the world had been obtained from this State, weighing 28 pounds.

Silver, *Lead* and *Zinc* have been mined to some extent for more than 30 years in middle region, chiefly in Davidson county, at Silver Hill and the neighboring mines, and recently they have been discovered in several of the western counties.

Copper has been found in more than a dozen counties, and a large number of mines have been opened in the last 20 years throughout the middle and mountain regions, and were wrought quite extensively before the war. Four or five of them have been recently re-opened, and put in operation on a large scale. It occurs in rocks of both Laurentian and Huronian age, chiefly in a gangue of quartz, but also in hornblende slate, syenite and tremolite, and in talcose slates. Most of the gold veins of the State contain copper in large part, and some of the mines of copper were first opened as gold mines. It exists mostly in the form of copper pyrites, although the other common ores are of frequent occurrence.

Mica. A great many mines of this mineral have been opened in the last 3 years, in some of the western counties of the State, most of them in Mitchell county, and all of them in the Laurentian rocks. It is found in ledges (veins) of very coarse granite. Many of the plates of Mica are of remarkable size, reaching 3 and even 4 feet in diameter. It is mostly sold in Philadelphia and used chiefly in the manufacture of stoves, and the mining of it is a very profitable and rapidly growing industry.

Graphite is very abundant in the State, both in the middle and west, existing chiefly in large bedded veins, generally more or less earthy and slaty, but occasionally quite pure and crystalline. It has been wrought on a large scale at several points. One vein, a few miles from the Capital, is one of the most extensive known.

Corundum has been found in large quantities in several counties west of the Blue Ridge, and is now extensively mined. Several valuable rubies and sapphires have been already obtained, one piece having been sold in London for $4,000: and a ruby crystal of 312 pounds is in the cabinet of Prof. Shepard, Amherst College, Mass. The principal use of this mineral however, is in the manufacture of the finer kinds of emery, for which purpose it has no equal.

Chromic Iron is of common occurrence in the same region.

Manganese. Several veins of the Black Oxide, of considerable extent have been found.

Barytes is found in large veins in the western and middle counties, and is exported to the Northern States, to be used, among other things, in the manufacture of paints as a substitute, in part or whole, for the lead carbonate.

Building Stones, granite, marble and sandstone abound everywhere.

Half a dozen *Diamonds* have been found accidentally in washing gold, some of them of considerable value.

Oil Shales exist in great thickness in connection with the coal beds and yield a large per centage of oil.

The other minerals mentioned are of common occurrence.

Of minerological curiosities there is a larger number found in this State than in any other of the United States.

CLIMATE.

The climate of North Carolina corresponds to that of Northern Italy and Southern and western France being tempered on one side by the Atlantic ocean and on the other by the high peaks and table lands of the Appalachian mountains. And as the State has so great a length from east to west, as well as so considerable an elevation towards the interior (3,000 and 4,000 feet,) the range of climate is very great, from *subtropical* on the coast, within the influence of the Gulf Stream, to *cold temperate* on the tablelands of the west. The isothermal in the one case, (at Smithville, the extreme southeast,) being 66° (that of Alexandria, in Egypt,) and in the other (at Boone, the higher mountain plateau in the west,) about 51°, which is that of New York and of Paris, France; the middle region falling under the line of 60°, which is that of Nagasaki, Athens, Gibraltar, &c. The following tables of temperature, taken from the U. S. Agricultural Reports and the Smithsonian Observations as given by Blodgett, will show the range and character of the climate better than any description.

Annual temperature,	59° Far.
Summer temperature,	75
Winter temperature,	43
Rainfall,	45 inches.

RALEIGH, N. C.	FLORENCE, ITALY.
60°	59°
76	75
44	44

BEAUFORT, N. C., (on the coast,)	GENOA, ITALY.
62°	61°
78	75
46	47

ASHVILLE, N. C., (In the mountains.)	VENICE, ITALY.	BORDEAUX, FRANCE.
54°	55°	57°
71	73	71
38	38	43

SMITHVILLE, N. C. (Sea Coast.)	MOBILE, ALA.	NICOLOSI, SICILY.
66°	66°	64°
80	79	79
51	52	51

Thus it will be seen that the range of climate in the State is the same as that from the Gulf of Mexico to New York. The influence of this circumstance is seen in the wide range of natural and agricultural products, from the Palmetto and Magnolia grandiflora to the White Pine, Hemlock and Balsam Fir, and from the sugar cane and rice to Canadian oats and buckwheat.

And while the cold of winter is not severe, 10° (of F.) being rarely passed, except on the highest plateaus, the temperature of midsummer is not so excessive or trying as further North, in New York, for example. While there are hundreds of fatal cases of sunstroke every summer in New York and other Northern cities, the disease is almost unknown in North Carolina. And last winter was one of unusual severity in the northern and western States, the thermometer several times dropping to 30° and 35°, and even 40° below zero in Iowa, Michigan and New York; but here 10° *above* zero was reached but once, and then but for a single night.

Healthfulness. Malarial diseases are of frequent occurrence in summer in the champaign country of the east and a hundred miles inland, especially along the river courses; not of a malignant type, however. But the middle and mountain sections are remarkably salubrious, with the exception of a few restricted localities on sluggish streams, just as in Iowa and on the upper Missouri. By reference to the sanitary department of the Census Report of 1870, it will be seen that one of the two or three most healthy localities in the United States is found in the western part of North Carolina, just east of the Blue Ridge. And indeed it would be difficult to find a more pleasant or salubrious climate in the world than the whole mountain region of N. C.

FORESTS.

It will be seen from the United States Census table for 1870, that of its 50,000 square miles of territory, 40,000 square miles are still covered with forests. The range and variety of prevalent and characteristic species of growth being of course proportioned to those of the climate and soil, are very great. There are in fact three well marked and broadly distinguished forest regions, corresponding to and dependent upon the three geographical subdivisions, eastern, middle and western. And while the first section is characterized by a growth common in its prominent features to that in the Gulf States, as the long leaf pine, cypress, &c., the western or mountain section contains many species familiar in the White Mountains, and in New York. Among the most distinctive, abundant and valuable species are the *Pines, Oaks, Hickories, Cypress and Juniper.*

Pines are the predominant growth of the eastern section. There are eight species in the State, the most important being the *Longleaf,* (Pinus australis,) the *Yellow,* (Pinus mitis,) and the *White,* (Pinus strobus). The longleaf pine is found only in the eastern or sea coast region; the yellow pine abounds throughout the State; the white pine is limited to the higher mountain regions.

The Longleaf Pine is the predominant growth of the eastern section of the State, and occupies almost exclusively a a broad belt, quite across the State, and extending from near the coast more than a hundred miles into the interior, covering a territory of near 15,000 square miles. This is one of the most valuable of all trees, on account of the number and importance of the uses it subserves. It is shipped in the form of lumber for civil and naval architecture to all parts of the world, and is unequalled for these purposes, on account of its strength and durability. It furnishes the *naval stores* of commerce, known in all parts of the world; the forests of this State furnishing twice as much as all the other States together. From the rosin of this tree is made the rosin-oil of commerce, and this substance also supplies the Southern towns with gas.

The *Yellow Pine* furnishes an important building timber in all parts of the State.

The White Pine is confined to the spurs and plateaus of the mountain region, being found in great abundance in some counties, and of great size, 3 feet and more in diameter, and 100 to 150 feet high.

The other species are less widely distributed and less valuable, except the Pinus tæda, which, in the eastern section, sometimes attains a great size, and furnishes an excellent building and ship timber.

The Oaks rank with the pines in value, and excel them

in variety of uses, number of species and extent of distribution. While a single species of pine gives character to less than one-third of the forest area of the State, the oaks dominate not less than two-thirds. There are 20 species in the United States, all of them found in North Carolina, with possibly one insignificant exception. Among these the most important are

The White Oaks, of which there are several species, (the most valuable, Quercus alba, Q. obtusiloba, (Post Oak,) and Q. prinus,) forming extensive forests in all sections of the State. On account of its strength and durability and great abundance, its uses are important and manifold, both for domestic purposes and for export in the form of staves and ship timber. The shipyards of Liverpool are already seeking their material in the forests of middle North Carolina.

Several other species of oak are also of wide and varied use, chiefly the *Red Oak*, (Q. rubra,) *Black Oak*, (Q. tinctoria,) and *Willow Oak*, (Q. phellos,) which are abundant throughout the middle and western district, and often grow to a very great size, *Live Oak*, (Q. virens,) is found only in the seaboard region, whose value in ship-building is well known.

Hickory. Of this tree there are 9 species in North America, and 7 of them are found in this State, and three species in all parts of it, and in abundance, and often of great size. But little use has hitherto been made of this tree, except as fuel and for wagons and handles; but being one of the most dense, rigid, heavy and iron-like of our woods, it has recently come into great demand, and many large handle and spoke factories have been erected within a few years, whose products are shipped by millions to Europe, California, Australia and all mining countries especially. The forests of North Carolina will supply this world-wide demand for many years.

Walnut exists in two species, one, the common Black Walnut, (Juglans nigra,) throughout the State, but most abundantly in the middle district. It is a most valuable wood, being very compact, durable, free from attacks of insects, of a very fine dark brown color, and capable of a high polish. It is the most popular and universally used cabinet wood in the United States, but is so common in the middle and western sections of this State that large farms are fenced with it.

The Chestnut, (Castanea vesca,) is one of our largest forest trees, sometimes 10 feet in diameter and 80 to 90 feet high, found mostly and abundantly in the Piedmont and mountain regions of the State, where it is much esteemed and used for fencing on account of its great durability and facility of working. It is also valued for its abundant crop of fruit, which, with the acorns of the oaks, is the principal dependence of the hog-raisers of the mountain counties.

Poplar, (Liriodendron tulipifera,) is one of the largest and handsomest of our forest trees, and occurs in all parts of the State, attaining its greatest size in the mountains. It is much used for building and other domestic purposes as a substitute for pine, combining lightness and facility of working with rigidity and durability.

Cypress, (Taxodium distichum,) abounds in the swamps and lowlands of the east, forming the almost exclusive growth of several thousand square miles of territory. It grows to a great size, the wood is very light, durable and much used for the manufacture of shingles, which are exported in immense numbers to all the Northern Atlantic ports. It is also used for building purposes, and for staves and telegraph poles, water vessels, &c.

Juniper, or White Cedar, (Cupressus thyoides,) is found in the same region, though not so abundant, and is used for the same purposes as the cypress, especially for shingles and cooper work, for which it is even preferred to the latter.

Besides these are the Maple, (6 species,) Birch, (3 species,) Beech, Ash, (4 species,) Poplar, (3 species,) Elm, (3 species,) Mulberry, Sassafras, Gum, (4 species,) Dogwood, Persimmon

Holly, Locust, (2 species,) Sycamore, Linn, (Linden or Lime, 3 species,) Buckeye, (2 species,) Wild Cherry, Red Cedar, White Cedar, Magnolia, (7 species,) Willow, (4 species,) and others, of various uses in domestic economy; many of them valued as shade and ornamental trees, a number of them much prized as

Cabinet Woods; among which may be mentioned the *Black Walnut*, already described, the *Red Cedar*, sometimes nearly equalling the Mahogany in beauty of color and grain, free from insects and aromatic; the *Black Birch* or *Mountain Mahogany* and *Wild Cherry*, both of very ornamental grain, taking a high polish; and so also the *Curly* and *Bird's Eye Maple;* the *Holly*, a beautiful, close-grained, white wood, taking a brilliant polish. It will readily be imagined what variety, richness and beauty these numerous species, belonging to so many and widely differing families of plants, must impart to the forests of this State, and what a vast mine of wealth they must become in the near future.

THE SOILS

Of the eastern section are generally sandy and of moderate fertility, (with occasional ridges very sandy and sterile); but along the streams are wide *bottoms*, and stretching out many miles from the bays and sounds, immense level tracts of clayey loam of great depth and fertility, producing 20 to 30 bushels of wheat, or a bale of cotton to the acre. And on the flattish swells, between the mouths of the great rivers, and around the margins of the lakes are vast tracts of swamp lands, covered with dense forests, of a dark peaty soil of great depth and inexhaustible fertility, producing the largest crops for 100 years in succession without manure.

In the middle and western districts, the region of predominant oak growth, the soils are of every variety of composition, and every grade of fertility. They may be generally described as clayey and gravelly loams, except the *river bottoms,* which are clayey, and sandy loams. As these soils are for the most part derived from underlying granitic rocks by simple chemical decomposition, they are arranged in parallel northeasterly zones of fertile and poor soils, but all capable of indefinite improvement by the hand of intelligent husbandry. This great variety of soils, together with the wide range of climatic conditions, gives rise to the greatest variety of natural products, and lays the foundation for an immense range of agricultural productions. One remarkable feature of the mountain section is that the highest ranges and peaks are covered with soil, and heavy forests crown their highest summits and steepest declivities.

POPULATION.

North Carolina came out of the Revolution of 1776 with a population of nearly 400,000. The settlers came originally from England, Scotland, the North of Ireland and Holland, many of them by way of Pennsylvania and other Northern States. Since the Revolution there has been less influx of population from other States than efflux Southwestward to the new States, and very little immigration from Europe. Being the third in population among the original 13 States, she is now 14th among 37, numbering in 1870,

White,	678,670
Colored,	392,891
Total,	1,071,361

This number is about 50,000 less than it would have been but for the late war, judging from the previous rate of increase.

INDUSTRIES.

Agriculture has always been the chief and almost exclusive occupation of the people of North Carolina—agriculture and the few simple manufactures and other occupations subsidiary to it. But within a generation several branches of *Manufactures* have grown up to some importance, especially those of *Turpentine, Cotton, Tobacco, Iron* and *Lumber. Mining* has also received a considerable impulse within the last 50 years, especially in *Iron, Gold, Copper* and (to some extent) *Coal*

AGRICULTURE.

The habits of the people have been agricultural from the beginning—land being very cheap and the population scattered, the great marts and markets of the world, until recently, distant and difficult of access, the principal forms of agriculture very profitable, and the conditions of climate and soil being favorable for the production of a great variety of crops, almost every crop grown in the United States being produced here, as may be seen from the columns of the Census Reports. Half of the territory of the State is adapted to the culture of

Cotton, of which in 1870, nearly 150,000 bales were produced, and last year probably 175,000, worth more than $13,000,000, and the number will nearly reach 200,000 this year.

Tobacco is the principal market crop of at least one-third of the State, more than half of the middle and all of the western district. The product as given by the Census of 1870, is still small compared with what it was before the war, but is extending very rapidly, and the quality is very high, the soils of more than a dozen counties, producing the highest grades in the United States, the product of a single acre frequently reaching 500 and $600.

The Grains, (Cereals,) of all latitudes flourish in one or another of the three districts. *Wheat* is an important crop in every section of the State, flourishing alike on the clayey champaigns of the extreme east and the high plateaus of the mountain section, being exported in considerable quantities from all the sections. *Oats, Rye, Barley* and *Buckwheat* of the best quality are produced in the mountain section, but their culture extends, especially that of the first two, through the middle and eastern sections.

The Grasses, (including Clover,) flourish best in the mountain region and are found here the in greatest variety and perfection; those most cultivated are, (besides Clover,) Timothy (Alopecurus pratensis,) Orchard Grass, (Dactylis glomerata,) Blue Grass, (Poa pratensis,) and Herds Grass, (Agrostis vulgaris). But Clover and Orchard Grass are cultivated successfully and easily throughout the middle region, and in proper soils also in the eastern counties. Native (wild) grasses abound in the eastern as well as the mountain region and form good natural pasturage, upon which in ordinary seasons, Cattle and Sheep keep in good condition throughout the year, so that one of the most profitable pursuits in either section is *Cattle and Sheep raising;* and there is no part of the State where these occupations are not profitable, whenever pursued intelligently; but the mountain district is pre-eminently adapted to this branch of husbandry and to *Dairy Farming* and *Cheese Making*. The latter industry has been introduced within a few years on the co-operative or New York plan, and has taken root in several counties. This region offers advantages over any other part of the continent, especially in the cheapness of land, favorable conditions of climate and proximity to the Atlantic seaboard and to the world's markets.

Grapes. The whole of this State is notably adapted to the culture of the grape and the manufacture of wine. The proof of this is, first, that a considerable number of the best American grapes originated within its territory, such as the Catawba, Lincoln, Isabella, Scuppernong, &c.; second, the testimony of the best observers and growers of the Ohio Valley, and of the whole country, and third and chiefly, the success of the few intelligent experiments that have been made. And this opinion is confirmed by the considerations of climate, which are demonstrably known to control this industry. In the remarks on climate it was shown that the larger part of this State corresponds in this important respect to middle and northern Italy, and to middle and southern France. On this subject Humbolt observes, (Cosmos,) "we find that in order to procure potable wine it is requisite that the mean annual heat should exceed 49°, that the winter temperature should be upward of 33°, and the mean summer temperature upward of 64°," and he cites Bordeaux, France, for which these figures are 57°, 43°, 71°: for this State, to repeat, they are 59°, 43°, 75°. When this subject shall be taken up in a practical, and intelligent and business-like way, by persons who understand its regulative conditions, there remains no room to doubt that this will become one of the leading industries of the State.

Apples, peaches, pears, &c., grow in all parts of the State; but the middle and western sections are among the finest fruit-growing regions of the continent.

The following abbreviated table from the United States Census of 1870, will give some idea of the range of agriculture production in the State.

Cotton,	144,935 bales.
Tobacco,	11,150,087 pounds.
Corn,	18,454,215 bushels.
Wheat,	2,859,879 "
Oats,	3,220,104 "
Rye,	352,006 "
Rice,	2,059,281 pounds.
Orchard Products,	$ 394,749
Wool,	799,667 pounds.
Peas and Beans,	532,749 bushels.
Potatoes,	738,883 "
Sweet Potatoes,	3,071,840 "
Molasses,	655,743 gallons.
Honey,	1,404,060 pounds.

The leading market crops, as seen from the table, are Cotton and Tobacco, and the culture of both is rapidly extending, and after them the bread crops, Maize and Wheat, are most largely cultivated, and to some extent exported. Potatoes are most abundant in the mountain section, where they are of the finest quality, and might be produced in immense quantities, and doubtless will be as soon as railroad facilities shall bring them better within the reach of markets. The Sweet Potato is a crop of great importance, especially in the eastern section, where it yields immensely, growing well even on thin soils. The crop of 1870 was just half that of 1860, and four times that of the Irish potato, whose place it supplies. Its chemical analysis, given in the U. S. Department of Agriculture Report, 1869, is, for a dried sample, as follows:

	SWEET POTATO.	IRISH POTATO.
Water,	7.65	7.95
Cellulose,	6.75	4.14
Starch,	65.29	63.47
Albumen,	1.21	11.70
Sugar,	14.83	7.57
Fat,	0.81	1.26
Ash,	3.15	3.91

The analysis of the Irish Potato given in a parallel column, shows the points of difference: the sweet potato containing more starch, cellulose and (especially) sugar, and the Irish exceeding in fat and especially albumen. It is very much used, and found very profitable for fattening hogs, and is latterly steamed and dried by machinery, and ground into flour for export. In addition to the products above set down as the most important, the Census tables show that a great many other articles are produced in considerable quantities, enough to show the adaptations of the climate and soil, if the energies of the people were not directed mainly to the production of two or three market crops. Among these may be enumerated Sugar, (both cane and maple,) Wax, Hops, Flax, Silk, Wine, Butter, Cheese, Barley and Buckwheat, previously mentioned, and Live Stock. The estimated value of annual farm products is $57,845,940.

MANUFACTORIES.

Agriculture being the leading occupation, as above stated, manufactures occupy an altogether secondary and subordinate place. But the facilities for many branches of manufacture are unsurpassed. Among the advantages may be mentioned first, the unlimited water power developed by the large rivers (previously described,) and their tributaries, in their flow from the mountains to the sea, through a descent of more than 1,000 feet; second, the abundance and wide distribution of fuel; third, the wide range and great abundance of raw materials, at hand, as Cotton, Tobacco, Lumber of all sorts, Iron ores and a great variety of farm products; fourth, the abundance and *cheapness of labor*, as compared with the Northern States; fifth, facilities for producing everything required by a manufacturing population, and sixth, a favoring climate—no obstructive ice. And as a matter of fact, those few capttalists who have embarked in enterprises of this sort find them very profitable; as for example, the cotton manufacturer, whose profits exceed 20 per cent.

The following list of manufactures will show that already some attention has been diverted from the production of cotton and tobacco to the more profitable business of converting these agricultural products into more valuable forms.

KIND.	NUMBER.	VALUE ANNUAL PRODUCTS.
Cotton,	36	$ 1,345,052
Tobacco,	110	718,765
Turpentine,	147	2,338,309
Lumber,	533	2,107,314
Fisheries,	42	265,813
Iron, Wool, Paper, Wood, Leather, &c.,		13,315,636
		$19,825,076

It will be observed by comparing the products of the different States, that in one article, Turpentine, (Naval Stores,) this State produces two-thirds of the total made in the United States.

Of course many of these annual products have been greatly reduced by the late war. And a large number of manufactures for domestic consumption are not specified—manufactures in wood, iron, leather, &c., as of buggies, wagons, agricultural implements, railroad machinery and rolling stock, shoes and many others.

MARKETS.

The distance to New York from the termini of the railroads and canals and from the navigable waters of the east, is about the same as that of the western part of New York State, and the facilities for reaching that mart are much greater—by railroad, by Chesapeake Bay and by the ocean route—lines of steamers from Norfolk and the other eastern ports. And there are lines of steamers from Norfolk to Europe direct.

PUBLIC SCHOOLS.

The State had a complete system of public education before the war and a school fund of $2,500,000. A new system has been inaugurated within two years, supported by taxation, and is gradually getting into working order again, and no doubt in a few years will be in full operation.

RELIGION.

The people of North Carolina are almost entirely Protestant, of various denominations; but all sects are equally free before the law and at the ballot, whether Protestant or Papist.

TAXATION

is very light, amounting this year to $50\frac{1}{3}$ cents on the hundred dollars, (including the school tax,) less than the half of one per cent; last year it was 40 cents.

THE PRICE OF LAND

varies with the distance from market and fertility. The price of average qualities is from 3 to 10 dollars per acre. Good cotton and tobacco soils may be had at these rates in large quantities; soils capable of producing, with intelligent culture, 100 to $500 per acre, and actually yielding at those rates in many cases.

The Swamp Lands of the east contain immense quantities of land of the highest fertility, requiring only drainage and clearing of timber to render it capable of producing a bale of cotton or 50 bushels of corn to the acre. Large bodies of these lands are owned by the Public School Board, and are held at $1 and less per acre.

Mountain lands are purchasable in large tracts at 50 cents to $1 per acre. These are good grazing lands and heavily timbered, much of them having a fertile soil, but mountainous, and for the most part admirably adapted to grazing purposes. The best quality of improved farming lands are 15 to $25, in exceptional cases selling for 40 and $50.

IMMIGRATION.

Several thousand immigrants have come into the State within the last two or three years, chiefly from Canada and the Northern States, and from Scotland and England, attracted by cheap lands and an admirable climate. There is a universal disposition among all classes to welcome immigrants from all quarters. And special arrangements can be made with the railroad companies, securing a reduction of fares and freights.

www.ingramcontent.com/pod-product-compliance
Lightning Source LLC
LaVergne TN
LVHW011143110826
845150LV00008B/2484

* 9 7 8 1 4 1 8 1 9 2 2 5 9 *